T0296555

GOOD PASTURES

THE OLD MONASTERY
converted into two farm cottages

GOOD PASTURES

Some Memories of Farming Fifty Years Ago

BY

T. BEDFORD FRANKLIN
M.A., F.R.S.E.

CAMBRIDGE
AT THE UNIVERSITY PRESS
1944

CAMBRIDGE
UNIVERSITY PRESS

University Printing House, Cambridge CB2 8BS, United Kingdom

Cambridge University Press is part of the University of Cambridge.

It furthers the University's mission by disseminating knowledge in the pursuit of
education, learning and research at the highest international levels of excellence.

www.cambridge.org
Information on this title: www.cambridge.org/9781107586635

© Cambridge University Press 1944

First published 1944
First paperback edition 2015

A catalogue record for this publication is available from the British Library

ISBN 978-1-107-58663-5 Paperback

Contents

Illustrations

Preface

My Father—C. H. B. Franklin—farmed Shutlanger Grove, Towcester, for 30 years. The following chapters are based on his notes and my own memories of the various operations described.

T. B. F.

May 1944

1. Introductory

FARMERS come and farmers go, but the land goes on for ever. The details of farming change with the times, but the underlying principles remain the same, and the wise farmer's creed, 'I believe in keeping the land in good heart', has not altered through the ages.

Father loved the land and everything to do with it; agriculture was in his blood, and he had the reputation of being one of the best farmers in his district. Grassland was his chief love and on it he lavished money and care, so that his Red Devons were well known and won numerous prizes at the local fat-stock shows, and butchers competed with each other to buy them.

He was a pioneer in many ways and used methods that were unique fifty years ago, but which now have become recognised as standard practice; he had not the advantages of modern implements and the new strains of grass seeds, but in spite of this his pastures were a picture and the admiration of all who came to see them.

There is a great difference in the attitude of mind of those who live off the soil and those who make their living on the soil. The former are well known in their district as land robbers, and are despised accordingly by all right-thinking countrymen. The latter would as soon rob the land as their mother, and when the time comes for them to give up their farm they do so happy in the knowledge that it is in better order than when they took it over.

On their broad shoulders lies the whole responsibility for keeping the land of Britain in good heart; so that fine cattle fatten in the pastures and good corn crops ripen to harvest on the ploughlands, and this, only too often, with but little profit to themselves after much hard work.

These memories, of good farming fifty years ago, are a tribute to this gallant company, of which, in his day, Father was a shining light.

2. *A Very Little Geology*

IT is often stated that geology is the real basis of all agriculture. This is probably true enough, but geologists and farmers are apt to look at the subject from very different view-points. The geologist is mainly interested in the rocks beneath the surface, but the farmer's interest lies in the soil at the surface.

There may be a direct relationship between the two, but in those parts of Britain that were covered with ice sheets and glaciers in the Ice Age the surface soil often has little or no relation to the underlying rock. For here the surface soil is not made from the weathered rock below, but from a mixture of many rocks carried long distances, ground down under the glaciers, and deposited when the ice melted.

It is in this ground-up mixture of many rocks that the farmer grows his crops, and it is the composition of this mixture that settles what crops he can grow best.

If nature did the grinding and mixing well, the farmer may have a good loam soil that will grow almost anything. If, as often happened, nature ground up the rocks a little too finely, the soil may be a heavy clay difficult to cultivate and best suited to grass. Or the grinding process may have been cut too short, when a hungry gravelly or sandy soil may result.

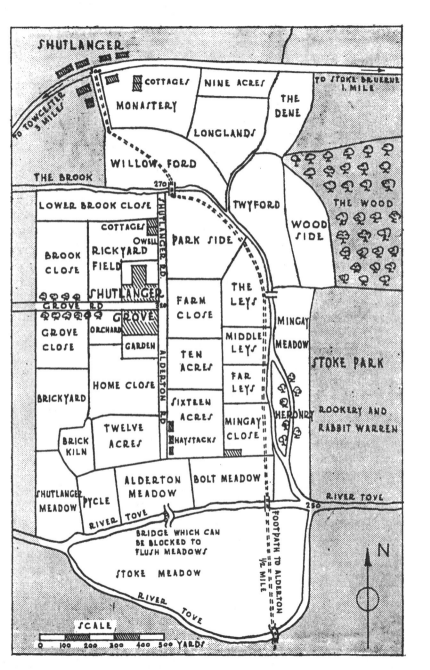

As soon as the ice sheets and glaciers melted streams began to run, and these carried with them the finer particles of the mixture and deposited them in the meadows near their banks. Subsequent winter floods increased the depth of these deposits, so that now many meadows have several feet of fertile alluvial soil on their surface.

Obviously the ground-up mixture deposited by the ice sheets and glaciers will be thickest on the lower ground and may hardly exist at all on the high ground. Here the soil will be directly related to the weathered underlying rock.

On the whole nature had been kind when the ice disappeared at Shutlanger at the end of the Ice Age. For, although the surface soil on the higher ground was thin and closely related to the sandstone and limestone underneath, yet, on the lower slopes and in the meadows, it was a deep fertile loam or good alluvial soil, which had no relation to the underlying rock as far as its composition was concerned.

But the underlying rock had a big effect on the drainage of the soil, and the behaviour of the good loam or alluvial soils was very different where it overlay clay or sandstone.

Father had taken the trouble to learn the geological history of his farm; he understood the reasons for the peculiarities of each of his fields, and was well rewarded in consequence. He certainly agreed that geology was a sound basis to agriculture.

3. The Country Scene

CLOSELY allied to the geology of the district is the appearance of the country scene. With us it was a chequer board of fields of different colours enclosed by quickset hedges studded with oak, ash or elm. Here and there hazel woods broke up the regular pattern, and lazy streams bordered by willow and sallow bent and twisted across the board.

The wood was a delight in spring with its carpet of bluebells, anemones and primroses, and a feast in autumn when the nuts were ripe. It was the reputed home of a badger, the undoubted home of many rabbits and foxes. Sometimes after dark a fox would visit us from his earth in the wood, and then pandemonium reigned amongst the guinea-fowl and hens that roosted on the beams of the covered yards; but they wisely only swore at him from the safety of their lofty perches, and he had to retire supperless and discomfited.

Perhaps the brooks gave us all the most pleasure, for along them there was always something to see at all seasons of the year. There we could watch herons busy fishing, standing statuesquely knee deep in the water until an unwary fish came within reach. Kingfishers darted up and down or sat poised like a diver on a tree stump; coots and moorhen and wild duck nested there, and occasionally an otter would take to the water almost noiselessly from the reeds.

Rosebay willow herb, forget-me-not, marsh marigolds and reed mace lined the banks with a blaze of colour in the summer.

There were fish to catch in the brooks—roach, dace, perch, eels and an occasional pike—when cooked they tasted of little but mud, but seemed excellent to those who had caught them.

Whether we went out bird-nesting, or with a camera, fishing rod, or dog and gun, the brook always had a fascination that never palled.

We collected wild ducks' eggs and hatched them under a hen and kept the young ducks in a wired-in enclosure in the orchard. Here they thrived and bred and became quite tame and had the run of a good-sized pond. We also kept young owls, taking them from their nests in hollow trees in the wood and risking the furious attacks of their parents. They too became quite tame and we fed them on rats and mice and moles. The rats and mice were mainly provided by our cat; she was a great ratter and always brought her catch to us to cash it for a piece of meat which she much preferred; the moles we caught ourselves from the fields shortly to be shut up for hay.

When the owls grew up we let them go, and they often returned in the evening bringing their friends with them for any food we had to give them; several owls sitting on the roof of the house hooting for their supper was a wonderful sight.

Father had a great dislike for rabbits, for they destroyed his grass, ruined his hedges and ditches, barked his young trees and nibbled his root crops; it was our job when we were old enough to carry a gun to keep down the rabbits on the farm. We shot a good many lying out in the grass all through the summer, and in the winter we ferreted them and filled up their holes. The farm workers set snares and caught a good many, for they knew that to Father any way of catching a rabbit was a good way.

We always had several coveys of partridges on the farm and usually got a bag of about ten brace at the first shoot in September. Our groom was a keen sports-man and always carried out what he called 'beating the bounds' in the early morning of the day of the first shoot. This consisted of going round the boundary fences cracking a whip to drive the partridges into the centre of the farm so that they could not so easily escape over the boundary before we had a shot at them. How much of our bag was due to his efforts we never knew, but we strongly suspected that in his zeal he beat the bounds well outside and not inside our boundary fences.

To find the wild flowers of the district the hedgerows and ditches had to be hunted; here, untouched by the plough or grazing animal, a process of natural selection had produced the flowers and weeds suited to the soil and which flourished best in competition with their neighbours.

High up in the hedgerow, vetches, goosegrass and bindweed climbed for a place in the sun. Lower down cow parsley, hemlock, mayweed, sow thistles, fat hen, burdock, knapweed and nettles struggled for the mastery with the ubiquitous couch grass. On the banks and ditch sides chickweed, speedwells, violets, prim-roses, cowslips and cuckoo-pints fought out their lowlier but just as intense battle for existence.

There was a large rookery on our neighbour's land, and rooks, starlings, fieldfares, plovers, and larks were our commonest birds on the grass fields and arable. A few pigeons led a somewhat harried existence, for we waged war on them at all seasons.

Except at sowing time and harvest the rook was not so black as he was painted, and but for him and the starling the arable fields would have suffered more from wireworms and leatherjackets than they did.

The eerie piping note of the plovers, the cawing of flights of rooks going home, and the noisy chattering of flocks of starlings going to bed in the trees, were familiar sounds to Father as he returned home in the twilight.

4. *Father and his Farm*

FATHER loved the land but he was by nature a grazier and not an arable farmer; grass was his abiding passion, and though he cultivated his arable land to the best of his ability, he did so as a means of providing the necessary winter food for his Red Devons, not for love of arable farming for itself.

He believed strongly that to get the best out of grassland it must be cultivated just as assiduously as the ploughlands, and he worked his grazing of it in rotation in much the same way as he arranged the rotation of crops on the arable. The upkeep of the fertility of the pastures was based on wise grazing by sheep and bullocks fed with cake, the use of farmyard manure, and supplementary dressings of lime and artificials when these were necessary. But he always contended that artificials must not be used without farmyard manure, and his version of an old saying ran:

> Lime and artificial without manure
> Will soon make farm and farmer poor.

Although he never doubted that a bag of artificial might produce an extra bag of wheat to the acre, he felt certain in his own mind that it was not the artificial, but the manurial value it released in the soil that produced the extra bag of wheat. This manurial value had to be replaced unless one was a land robber.

He always told everyone who asked his opinion on this vexed question, that he thought those farmers who used artificials without manure were running up an overdraft on the bank of fertility. They might cut a fine dash for a year or two, but the overdraft would have to be paid back with high interest in the end in the form of very large quantities of farmyard manure.

That a farmer should try to build up a good reputation seemed to him essential, and he held that the reputation must cover good relations with his workers, good-quality produce from the farm, good condition of the buildings, gates, fences, hedges, etc., and most of all good heart in the land. It would please him mightily could he know that, forty years after he left Shutlanger, there are still old men in the village who speak of him as the 'Master', and butchers in the towns who still talk of the wonderful fat beast they bought from him out of the big meadow.

Father carried out many experiments on various subjects, such as the best time of year to sow grass and clover seed mixtures, the best manurial treatment for producing extra hay, making silage without a silo, defeating wireworm and turnip fly, and the flooding of one of the meadows for an 'early bite' or hay crop. Though he deplored the lack of enterprise of many farmers he realised that most of them had to keep a tight purse, since they had so often learnt that argument and specious advertisement were only a device to make them 'buy a pup'. And he always remembered the incident, in the market-place at Northampton, of a seller of hair restorer who was shouting, 'This will grow hair on a pot pig, gentlemen, and only one shilling.' Just then a gust of wind removed his wig and revealed a completely bald pate. Shouts of derision greeted this mischance, and with the parting thrust, 'Try it on your own pot first, mister', the crowd melted away.

Small wonder perhaps that farmers generally were chary of trying out new schemes; our neighbours much preferred that Father should do the experiments so that they could copy his lead the next season if the result was a success.

Shutlanger Grove, ten miles south of Northampton, lent itself admirably to all these purposes; for it lay on the border-line between the best arable land to the east and the best pastures to the west, and so combined most of the good qualities of both.

Here was a farm of about 500 acres with great possibilities; in the fields on the higher ground, where the soil was a somewhat sandy loam, stock could winter out on the grass until Christmas and the arable fields would grow splendid crops of barley, oats or potatoes. On the lower slopes, where the soil contained more clay, it was a fertile loam suitable for wheat, roots, and good grass; while the meadows of alluvial soil on clay never dried out in drought and would carry a bullock to the acre all summer.

All the meadows, watered by three different streams, were outstanding permanent pastures; two of the largest, making up about 100 acres between them, were considered some of the finest pastures in the district.

About 120 acres of the farm were arable and this provided all the winter feed for the stock as well as some corn to sell.

There was an immense rickyard and barn with four huge bays, each big enough to hold a good-sized stack, and the doors on each side were large enough to take a fully loaded wagon. Covered ways connected the barn and granary to all the fattening yards, stables, and milking sheds, so that in bad weather in winter work could go on in comfort. Water was laid on from a well with a windmill pump some distance away from the buildings and was free from contamination by the manure in the yards.

Four good cottages on the farm itself, available for those who had to work on Sunday with the livestock, added greatly to the ease of working and the comfort and efficiency of the men. Two of these were close to the farm buildings and two near the village in the old Monastery. A beautiful porch and a spiral stair leading up to the Prior's room were all that remained of the original building, but clever restoration in keeping with the old part had resulted in two excellent and rather unusual cottages (see *Frontispiece*).

In fact, the whole place was admirably equipped and well suited by nature for fattening bullocks, and on this Father lavished money, time, and most of his energy.

But he also found time to do a lot of public work on the Assessment Committee and the Board of Guardians, and was Rector's warden at Stoke Bruerne Parish Church for all the years I can remember.

He hunted twice a week with the Grafton in the winter and so knew intimately all the Grafton country. From seeing what other farmers did with their land for miles around he added continually to his stock of farming lore, which was a veritable mine of information on all matters relating to the land.

He farmed Shutlanger Grove from 1875 to 1904, and then, on doctor's orders, he gave up the heavy responsibility of looking after the 800 acres he had at Shutlanger and Stoke Bruerne.

But he always insisted that he must have a little land wherever he lived, and first in Kent and finally in Sussex he went on fattening his beloved Devons.

In Sussex he bought a house and just 100 acres of land, and was considered an authority on all aspects of farming, and was the Central Landowners Association representative on the County Agricultural Committee.

He died in harness, having demonstrated good farming continuously in one place or another for over sixty years.

5. *The Farm Workers*

FATHER worked on the principle that no farmer would get his farming done well without the willing co-operation of his men; it was just as important to keep up the good heart of the men as of the land.

The best workers would only stay on a farm that was well equipped, efficiently run, and had a name for good-quality produce; though they might get the same wages on a poor farm they would never be happy there for they hated doing good work to no purpose. 'Good land, good heart, good work' appealed to them just as much as to the farmer; they would never work willingly for a land robber.

An unwritten law on a farm that everything must be done well encouraged the good countryman; such a farm soon collected a splendid team of men who took a pride in all they did for it.

Fifty years ago wages were low, but rents and the cost of living were low too. Our workers were a healthy and contented lot, they knew that if they or their wives and family were in trouble they could come to Father or Mother for the help they needed.

Mother often dealt with stiff backs and septic hands, took invalid food to sick wives or children, provided clothes for the new arrivals in the families, or a few shillings for rent in difficult times. It was a recognised custom for anyone in trouble to come up to 'The House' to have their difficulties smoothed out; with the result that the men respected Father as a good master, while their wives took Mother to their heart. We had many friends in the cottages in the village.

The village was the hub from which radiated good-will and mutual help—the oil which made the wheels turn and friction disappear.

If there was a vacancy for a man on the farm it was filled by a relation of one already there, we had four pairs of brothers working for us. Village boys began their working life on the farm and the girls helped in the house. The village, the farm, and the house were an integral whole whose well-being depended on the success of the farm.

Besides the regular workers from the village we had three Irishmen from Connemara who came every year in May and went back home in October.

Every spring, about the end of April, Father got a letter saying, 'James and Martin and me are ready to come if you will send the cost.' So the fare was sent and a few days later the three would turn up for the summer and take over once again the small house at the end of the cartshed.

Good workers, kind, and always willing to do any-thing for us, they thought nothing of walking three miles in the middle of the night to fetch the doctor in emergency. They got fuel and milk and potatoes from us and bought the rest of their food in the village; they were frugal and saving and took most of the money they earned to their wives and families when they went home.

Every year they brought Father a bottle of Irish whiskey, and when they went home again took something for their wives from Mother. How many years they were with us I cannot remember, but they stayed on until Father left Shutlanger; after that they would not return as they said they could not bear to work for a new master.

The standard of skill of our band of workers was high and Father never had to employ an outsider for any of the farm work; we had expert thatchers, hedgers and ditchers, drainers, carpenters, and masons on the spot. But skill at manual labour and knowledge of the art of feeding livestock do not always go together; cowmen are liable to give an unnecessary amount of cake to their favourite cows, cattlemen are often too lavish with food in the yards and too sparing on the grass. Many of them never realise that the beast get no exercise and good food in the yards, but a lot of walking up and down to graze their fill of poorer food when the grass fails in the summer.

One day Father found the following doggerel rhyme pinned up in the barn:

THE ART OF GRAZING

Little stalk, leafy mat,
Short walk, beast fat.
Much stalk, grass fine,
Long walk, beast pine.
Rests make clover feast.
Give cake, prize beast.

So our cattlemen did realise the importance of watching for the falling off of quality in a pasture, and the need of a rest or supplying a feed of cake when this falling off took place.

A bonus system for every foal, calf or lamb encouraged the wagoner, cowman and shepherd to spend extra time on the production of young stock, while the cattlemen reaped a similar reward when their bullocks won prizes at the shows. With these bonuses and a considerable amount of overtime at hay and corn harvest our men earned a good deal more in the year than their bare weekly wages.

The usual midday meal was a good slice of fat bacon between two thick slices of bread, cheese, and cold tea. This each man brought with him in a rush basket slung on his back when he came to work in the morning. During hay and corn harvest, when work went on late, the wives and children brought tea to the field and often stayed and had a family tea there.

The afternoon pilgrimage of the family from the village to the harvest field on a fine afternoon was one of the pleasant relaxations that the wives enjoyed. If it could be timed to fit the moment when the mower cut the last swath of grass or the binder cut the last width of standing corn so much the better, for then the families would gather round and help to hunt the rabbits that had been foolish enough to wait in cover too long, and many a bunny went home with the tea things.

One of the cattlemen had a splendid way of catching a rabbit lying out in a tuft of long grass in the pastures. Without any warning he would suddenly fall full length on the top of the tuft and then produce a very scared and somewhat flattened rabbit from the depths of his

stomach. He was so expert at this that I fancy his
family must have got tired of eating rabbit.

Holidays of a fortnight in the year were taken between hay and corn harvest when farm work was slack-
est. The Irishmen, of course, did not want a holiday and
took over from the wagoner, cowman, and shepherd
while they were away; the two cattlemen took their
holidays in turn, so that they did not have to delegate
their important work to a stranger just at the critical
time in the fattening period.

In this way each man got his holiday and the essential
work of the farm went on without loss of efficiency.

6. Weeds

So long as the wild flowers remain in the hedges and on
the roadsides the farmer is prepared to admire their
beauty, but as soon as they spread over the arable and
grassland they become weeds to be pursued to final
destruction with relentless energy and purpose. For
arable land foul with weeds and pastures full of everything but grass are patent evidence of bad farming and
must result in poor crops and badly fed livestock.

The successful eradication of weeds on a farm is no
simple matter; the methods used must vary with the
type of soil, the kind of weed, and the general response
of each individual field to the processes of cultivation.

The farmer must draw on his store of farming lore
to know what to do. He may have to smother the weeds
by a heavy crop in one field, kill them off by intensive
surface cultivation in another. In one pasture he may
get rid of them by judicious grazing, and by cutting
them down with the mower or scythe in another. Nor

do his difficulties end there; for a treatment that is fatal to one weed may be an encouragement to another, or detrimental to the texture of the soil. Lime will do away with corn marigold, knawel, sorrel, spurrey and runch, but will encourage charlock and poppies. Kainit is particularly effective in subduing charlock, but tends to make a heavy soil sticky; lime will restore the texture of the soil, but encourages charlock for the next season.

On our arable land the continuous growing of cereal crops promoted couch, charlock, poppy, corn marigold, buttercup and bindweed. These could be killed off by the intensive cultivation necessary before sowing a root crop, and by the hoeing and singling given to that crop late in summer; but if more than one root crop was grown in succession it encouraged the appearance of fat hen, small nettle, groundsel and fumitory.

So a rotation of crops was necessary, not only for the sake of freedom from disease and better yield, but for preventing any one set of weeds from getting too firm a hold of the field.

With us barley was the best cereal smother crop, especially if a full allowance of seed was sown; if, therefore, it was used as a nurse crop for a grass and clover seed mixture it had to be sown very thinly or a very poor strike of the seeds resulted.

For this reason Father preferred a thinly seeded crop of short straw oats, cut green before harvest, as his nurse crop, although this was not the custom of the district.

If a field got really foul with weeds a summer fallow might be necessary. By shallow cultivation every few weeks the weed seeds were encouraged to germinate and then destroyed; and those weeds with creeping rootstocks, by having their tops cut off continually, were

obliged to draw so heavily on the food in their roots that
they gradually died out.

Weeds in the pastures were even more of a problem,
for here the different preferences and different methods
of grazing of various kinds of stock came into action.
Fields grazed entirely by bullocks became infested with
ragwort, oxeye daisy, sorrel and buttercup, but when
grazed by sheep these weeds hardly ever became a
nuisance.

Overgrazing a pasture without proper rests produced
large crops of daisy, creeping buttercup, plantain, rib-
wort and silver weed; these appeared at once on any
bare place due to overgrazing or excessive treading
Undergrazing was no remedy, for thistles, ragwort and
yarrow always sprang up in profusion on the rough un-
grazed parts and had to be destroyed at intervals by
hand pulling or cutting over with the scythe before they
seeded.

In general, Father found that the best way of keeping
his pastures weed free was to keep a good sward by
suitable manuring and cultivation. He backed this up
with controlled grazing of both bullocks and sheep,
with short rests to enable the grasses and clovers to
recover and keep the weeds smothered. But there was
no royal road to success in all years either on the grass
or arable; for the weather had a great effect on the weed
population as well as on the virility of the grasses and
clovers, and no two seasons were exactly alike.

No wonder Father always considered the solution of
the weed problem as the acid test of a good farmer's
ingenuity and doggedness. He himself was remarkably
successful in keeping his farm weed free, and some of
his methods will be given in more detail in later chapters.

The freedom of his pastures from ragwort, docks, thistles and nettles was in no small measure due to the destruction he did amongst them with the stout ash stick that he always carried on his walks round the fields. He broke so many sticks in this way that we were never in doubt what to choose when we wanted to give him a present.

7. *Making a New Pasture*

To break a pasture makes a man,
To make a pasture breaks a man.

THIS old saying was full of meaning to Father, for he knew the wonderful crops that could be got when an old pasture was ploughed up and the accumulated fertility of years of grazing and manuring was liberated and cashed. But he knew too that many a farmer skimped the preparations for making a new pasture, and after all his labour and expense often reaped little reward. To him a pasture was something on which bullocks would fatten for years to come, and its making was a serious matter involving many operations and much time before the results would be fully seen.

Father always thought that calling a pasture permanent was a mistake; for though the field might be under permanent grass it would not remain a permanent pasture. The natural tendency of all pastures was to lose clovers and the more nutritious and palatable grasses, and to replace these by less valuable grasses and weeds with a consequent loss of fattening power.

When the decision was made that a pasture was beyond patching up and that the plough must go in, a long train of operations was set in action that would culminate in a new pasture some two years hence.

First of all the ditches were cleared out and the drainage examined to ensure that it was in good working order, and the hedges trimmed or laid as might be necessary. Then a dressing of farmyard manure and the dirt from the ditches was given to the field, after which it was ploughed and given a dressing of lime. This was all finished by October, and the field lay fallow all winter until the following April, when it was sown with lucerne under a nurse crop of about half the usual crop of oats.

The oats were cut green as soon as they came into ear and used as fodder. All that autumn the lucerne pushed down its long tap roots into the soil and so opened it up and aerated it; at the same time it was fixing nitrogen from the air in the soil. Lucerne grew very well and quickly with us, and by the following summer there was a grand crop of fodder for the horses and green stuff for the silage stack. The aftermath was grazed by a large number of sheep folded on it and fed on cake, to give additional manure ready for the grass and clover still to come.

The plough followed up the folded sheep in October, and after several cultivations the field was ready in the following April for the nurse crop of oats and the seeds mixture.

Father did not follow custom blindly in sowing his seeds in spring under a nurse crop. He had tried spring sowing without a nurse crop and found that he got a poor strike of seeds unless the weather was kind. He also tried sowing in autumn with even worse results, especially with the clovers, if the weather was wet and cold. Experience showed that with our soil and climate the safest way was to sow in April under a half-nurse crop, which was not left on till harvest but cut green in July.

So one fine day in April a whole battery of implements followed each other across the field. First the corn drill sowed about half a crop of oats followed by the mineral distributor which put on 8 cwt. of basic slag to the acre, then came the ribbed roller to make the small furrows into which the seed barrow dropped the mixture of grass seeds and clovers. The seeds were covered by a light chain harrow and firmed in with a flat roller.

This was an imposing cavalcade of horses and implements, which looked very workmanlike in the April sunshine, and gave Father a feeling of satisfaction at a job well done at long last. By evening it was all over, and at dawn the next day the 'bird hollerers' went on duty to keep the rooks, pigeons, and sparrows from undoing any of the good work. Woe betide them if they went to sleep at their posts during the 10–14 days before the seeds were safe, for this was now Father's pet field.

Father made up his own seed mixtures, for he thought he was more likely to know what suited his land than any seed merchant. The mixture for this field was as follows:

Red clover . . .	4
White clover . . .	4
Perennial rye grass . .	10
Cocksoot	6
Timothy	4
Foxtail	4
Rough-stalked meadow grass	4
	36 lb. per acre

Clovers always did very well with us. White clover spread and the pastures were full of it; even red clover was persistent, and in some fields spread almost as readily as the white clover. They were both good drought resisters, and not only fed themselves but the grasses round with the excess nitrogen they fixed from the air.

Clovers and rye grass grew well together, and the rye grass made a grand bottom grass and checked weeds. It certainly made a heavy demand on the soil to produce its big and nutritious crop, but the animals liked it and grazed it heavily. In spite of this it kept on growing and made an excellent grass for permanent pastures.

Cocksfoot gave the bulk so needed by bullocks which grazed it readily; but it was most persistent, and stood drought well thanks to its deep roots.

Timothy was a most useful grass with us, as it gave good grazing late in the season; it also tillered well and was most persistent.

Foxtail was first class for the 'early bite', and was much in evidence in March and April; the animals liked it, but in spite of heavy grazing it persisted and was a most valuable pasture grass.

Rough-stalked meadow grass was the real standby in winter and spring; it made a good thick bottom, checked weeds, and stood up to grazing without dying out.

The thin crop of oats covering the seeds was cut green in July and used as fodder or put on the silage stack. After that the young seeds were lightly grazed by a large number of sheep for short periods at a time, with some weeks rest in between, until the wet weather came and brought an enforced rest until the spring.

In the following year care had to be taken to ensure that the grazing was sufficient to prevent the grasses and clovers seeding; but on no account must the ground be poached, and it was usual to take off all the animals in wet weather.

But in good weather the treading of sheep and bullocks was the best form of cultivation and encouraged the grasses to tiller and the clovers to spread. By the third year after seeding the new pasture took its place as one of the permanent pastures and could hardly be distinguished from them.

This was Father's method of producing a new pasture; it was certainly elaborate, but those who sometimes mocked during the long preparations came back in later years to admire the results.

There was a good deal of talk fifty years ago of the virtue of sowing certain herbs—burnet, chicory, plantain, etc.—with a grass mixture, on the assumption that animals tired of the rather restricted number of grasses found in most fields, and so grazed these herbs readily. This idea never appealed to Father, although he knew that on plenty of poor grass fields in our neighbourhood, which were deficient in lime, potash, and phosphate, the animals certainly did graze these herbs readily. But there was never any preferential grazing of them shown on our own good pastures which were properly manured. To him the whole argument smacked of excuses for a poor pasture rather than a plan to improve a good one.

Although he had not at his command the splendid new strains of seeds available now, I do not think that the modern grass farmer could have taught Father anything about the sowing and management of a pasture.

His success with his animals was, I believe, due to his intense interest in everything to do with grass, and to the fact that he was not afraid to experiment.

8. *Renovating an Old Pasture*

THE peaceful pastures of the artist, poet, and writer are fantasies of the imagination.

Every pasture is the scene of a constant battle for the survival of the fittest, and though the battle is noiseless it is none the less deadly. Thoroughbred grasses, like thoroughbred horses, are 'kittle cattle'; without constant care and attention they will go down before the onslaughts of the commoner strains, which by virtue of their rude stamina will crowd out and smother their weaker brethren.

Though the grass remains the strains alter continually for the worse and the quality goes down until one day the farmer realises that what was once a good pasture is now but a poor thing and will no longer fatten anything like a bullock to the acre.

Father always made a rough count of grasses and weeds every time he shut up a pasture for hay; by this means he had a rough three-yearly check on every one of his fields, and could give suitable help and encouragement to the failing strains before their conditions were so desperate that ordinary remedies were no use.

When he took over a grass field adjoining the farm from a neighbour one September I think he did so because he knew it was on its last legs, and he wanted to try his hand at renovating it without ploughing it up. At any rate there was the light of battle in his eye when he and I went to inspect it thoroughly and decide on what course of action to take, and he enjoyed writing down its good and bad points.

Its good points were but two: a gentle slope to the south and a few remaining patches of white clover and rye grass; its bad points were many. The ditch at the bottom was blocked up completely and the drains had long ceased to run; the hedges were thin at the bottom, full of brambles, and ten feet high; the top of the field was covered with ant heaps, the bottom was boggy and full of reeds; one hedge was riddled with rabbit burrows and the rabbits had destroyed most of the grass near it; the whole was full of sorrel, matted with bent grass and weeds and obviously sour.

We decided to tackle the rabbits first, so had a morning's ferreting with two guns and killed quite a number. Then the burrows were filled with Renardine and blocked up with earth and turf. A fortnight later only two burrows had been reopened and ferrets were put in again, but this time no rabbits bolted and we dug out three more. After that there was no further sign of them, they were either all killed or had gone elsewhere.

The ditch was tackled next; it was deepened, the drains unblocked, and the earth so got was heaped ready for use on the field later; at the same time the ant heaps were cut off and added to this heap.

A dressing of about 20 cwt. of lime to the acre was now broadcast with a shovel over the field; this was intended to loosen and rot down the mat of bent and weeds, cure the sourness, and generally facilitate the cultural operations to be done in the spring.

Next the hedges were tackled and cleaned of brambles and undergrowth and then laid, so that by next spring they would be fit to keep in any stock and look more like good Northamptonshire hedges.

This done, the field was left for the winter; but later on, when the rain came, we noticed that it took quite a long time before the drains cleared themselves of the brown foul-smelling liquid that came away at first. This was a sure sign that formation of humus had stopped for a long time and that improperly digested resources of manure were coming away.

In the spring Father put a well-weighted sharp-toothed harrow into the field and dragged out most of the mat which now came away fairly easily thanks to the lime. The field was harrowed up and down and across and the mat collected into heaps. Then in one corner of the field he had a compost heap made of alternate layers of earth from the ditch and the ant heaps, the mat, some manure he had drawn there for the purpose, and a sprinkling of lime.

After the complete scarifying with the harrow the field looked pretty poor and ragged; so it was given a dressing of 10 cwt. basic slag to the acre, chain harrowed and rolled and then left to itself for several weeks.

When the grass began to grow it was lightly grazed on and off all summer by sheep in the same way as a new pasture, and in the autumn the manure heap, now well rotted down, was spread over it and the sheep taken off for the winter.

It was in the following spring that an almost miraculous transformation took place in the sward. White clover, birdsfoot trefoil, and rye grass appeared as if by magic from nowhere; sorrel, bent grass, weeds and reeds were all on the wane.

After another year of carefully managed grazing, the

old tumble-down pasture was taking its place amongst
the fattening pastures of the farm.

This job gave Father enormous satisfaction; and
years after, when the beauty treatment of face lifting
came into vogue, he often used to refer with pride to
our 'lifting the face of old *N*'s field'.

9. *Stocking the Larder*

For the fly, the fly, the fly be on the turmut,
And it's all me eye for me to try
To keep fly off the turmut.

So ran the song, and Father always hoped that April
might be true to form and give plenty of showers to
make the turnips grow quickly and keep the fly off them
until they had got their second pair of leaves. For in
dry weather, when the turnips grew slowly, a good
plant all over the field might be eaten off by the fly in
a day and the crop had to be resown.

Turnips, swedes, and kohlrabi were all subject to this
disaster, but mangolds were nothing like so liable to
mischance and so were frequently grown.

Father made many experiments to try to get the
better of the fly and found eventually that broadcasting
a mixture of salt, lime, and soot was as effective as
anything. It was a most unpopular job with the broad-
caster but it seemed to give good protection, probably
due to the absorption of moisture by the salt in the damp
evenings and mornings.

Root crops were 'ornery' as the saying went with us;
ornery had many meanings—awkward, cussed, troublesome, a nuisance—and it was a good word for them. For

the sowing was a more awkward business than with corn, the prospect of a plant was uncertain, the labour of hoeing and singling was troublesome in bad weather and might last well through haymaking, and work never ceased in the root field until the leaves met across the 24-inch rows sometime in July. Even then rabbits, hares, and pheasants took their toll, and mangolds were certainly susceptible to frost.

With us turnips and mangolds got different treatment; for turnips needed plenty of surface moisture and so were grown on the flat, but mangolds with their long tap roots were able to stand drought and were grown on ridges.

For both of them an enormous amount of cultivation and heavy dressings of manure were needed, and both demanded a very fine tilth to ensure a good plant. Anything up to 20 tons of farmyard manure to the acre ploughed in, 3 cwt. nitrate of soda and 3 cwt. basic slag per acre sown with the seed, and 2 cwt. per acre of salt after singling was Father's idea of feeding mangolds. It seemed a lavish allowance; but they were a very responsive crop, and some at least of the dressings would help the following wheat crop.

As soon as there was a plant horse hoeing was done, then hand hoeing down the rows until the mangolds were about 20 inches apart. A mangold seed really contains several seeds in one case, and so the young plants tend to be bunchy and a good deal of singling had to be done also.

At the end of all this the field looked a wreck and at first glance all the young mangolds seemed to have gone. But a broadcast dressing of salt and a good shower of rain put everything right, and the chosen plants grew away fast, while the rejected ones and the weeds died in the furrows between the ridges.

When the weeds were dead the horse hoe was used again to throw back some of the soil on to the ridges, and then at last Father heaved a sigh of relief that his mangold crop was well on the way to harvest.

But a lot more work had to be done before the mangolds were safely in the larder; and in the autumn before the first hard frosts came the whole massive weight of them had to be lifted, topped by hand, and carted to the rickyard. There they were stored in a clamp and covered with straw and earth.

This manhandling of a heavy crop was hard and slow work but was unavoidable; for bruised or broken mangolds went rotten in the clamp and infected many others, so that the loss might be considerable if the job was not properly done.

Most farmers hated haytime and were often very short tempered in June, but Father viewed it with equanimity. He did not mind very much whether he made hay or silage, and in fact he always made some of both. If June was fine more hay and less silage was made, if it was wet more silage and less hay, that was the routine.

Fifty years ago silage was something of a novelty, and many farmers were put off making it because they thought special arrangements were necessary and the result was a good deal dependent on chance. But Father made silage much as he made hay, the only differences being that only a little grass was added to the stack at a time, a layer of salt was sprinkled on every time the stack rose a few feet, and finally the stack was well weighted down with railway sleepers and stones on the top.

Of course the cowman and cattleman were extremely sceptical of any good result the first time it was made. 'The cows woant touch it, master, it'll be just like dung', was their comment; but they lived to take back their words when it was found in the following winter that cows and bullocks preferred it to good meadow hay when given the choice of both. And no other dire disaster, such as 'Milk'll turn sour on that stuff', ever resulted, for the milk yield was increased and the bullocks fattened well on silage.

Father never built a silo or ever found any need to do so to produce good silage which was readily eaten by all stock. But experience showed that not more grass should be cut than could be put on the stack in a few hours; for silage should be made of green grass and not half-made hay, and the grass should be cut before it is as ripe as for hay.

It was heavy work handling the weight of green grass, but the only complaint Father ever made was that the first year he made silage he lost his touch for knowing when hay was fit for stacking, and for the only time in his life had a hay stack that heated.

In really good weather of course we made hay. The two-horse mower cut down the grass, and as soon as the swaths were dry on the top the tedding machine threw them into the air and scattered them. When the scattered hay was nearly made the horse rake collected it into rows ready for loading on the wagons.

Such made hay as could not be carried in the day was piled up into haycocks in case of rain in the night. In this way it came to no harm though it had to be opened out again next morning, as soon as the sun shone, before being carried.

As most of the hay was for fattening quality bullocks, good quality rather than quantity was Father's aim at haymaking time. 'One inch at the bottom is worth four at the top', and we always cut our hay much earlier than our neighbours to avoid the inches at the top as much as possible.

We made a good many trials to see what treatment gave the most increase of hay, and found that on good pastures, which had been properly grazed and manured, an autumn dressing of 5 cwt. of basic slag to the acre would produce on the average about 25 per cent increase in the hay crop. On the poorer pastures this dressing hardly produced any increase of hay, although the improvement in the white clover was most marked. On these pastures a dressing of farmyard manure in the autumn gave a much better hay crop.

On wet days, or when there was no hay ready for carrying, a little silage was made; and then the men went down to the brooks to cut hazel, sallow, and willow spars ready for the thatcher, or pulled straw and laid it straight in bundles for the thatch. For as soon as the hay ricks were finished the thatcher began to work on them to prevent any loss of good hay and to get them done before corn harvest began.

We had an excellent thatcher amongst the men, for as Northamptonshire had many villages where the cottages were mainly thatched the art of thatching was still passed on from father to son in a good many families.

Thatching a rick was done in just the same way as slating a house; the work began at the eaves and each layer of straw was pinned to the rick by the supple spars of hazel, etc., which had been soaked in water to make them bend over easily in the form of a staple without breaking. The staples were then joined by lines of binder twine to keep the straw down tight, and as soon as one layer of straw was on another was laid on the top of it and overlapping it just as the tiles do.

So the work went on until the ridge was reached, when one layer of straw was laid over both sides of the ridge just as the coping tiles were put on the slated house. When all was done some reinforcement with more staples and twine were made at the corners to prevent the wind from blowing up the thatch at the vulnerable spots.

The thatch was then raked to pull out loose straws, the edges of the eaves were clipped, the thatcher's sign manual of some bird or beast in straw was put at one end of the ridge, and the job was done.

Once the straw, spars, twine, and a long ladder had been collected and delivered at the rick, one thatcher, and a boy to carry for him, did the whole thing; no wonder thatchers took a pride in doing such good-looking and useful individual work.

It was essential to have more hay and silage in hand than would be used by the stock during the coming winter; for the bullock loomed so large in the business of keeping up the fertility of the land that any reduction in their number in dry years when grass was short must be avoided at all costs.

Unless the number of bullocks was reduced hay or silage must be given to keep up the fattening process when the grass fails; selling beast at difficult times was always poor business, for all the farmers in the district would be in the same difficulty and no one would want to buy half-fat beast when their own grass was scarce.

CHARLES HENRY BLUNT FRANKLIN
1904

The only remedy was to keep up one's stock, and by making extra hay and silage in the good grass years to ensure having a stack or two in hand to tide over the lean years. This was Father's method; he always had some stacks ready for such an emergency, and thus equipped could carry on without worry whatever the weather during the summer grazing period.

10. Getting the Bedding Ready

FATHER could not have produced fat bullocks in the way he did without his cereal crops; for he needed oats for his horses, oat straw and middlings for fodder, and wheat straw for bedding for his animals wintering in the yards.

He did at one time give a good trial to peat moss litter, but it proved no better than straw; it was disliked by the men, was expensive, and had a poor manurial value.

So the bedding for the winter was first thought of fifteen months previously when the plough followed up the sheep folded on the turnips and winter wheat was sown.

In spite of the efforts of the 'bird hollerer' some of the wheat was eaten by sparrows and rooks, and in the following spring the harrow had to loosen the hard crust around the wheat plants formed by the winter rain, and this and the horse's hooves destroyed some more of the young plants. But the rest tillered better after the treatment and the loss was not noticed; a root that tillered well threw up many heads instead of one, and this must happen for a bumper crop of wheat and straw.

Owing to the various losses much more seed has to be sown than grows on to harvest and the Northamptonshire saying

One for the sparrow,
One for the crow,
One for the harrow,
And one to grow,

expresses all this in few words.

But the rhyme left out the wireworm, which was useful enough in the grass fields but a pest in the arable land. One year we had a good field of winter wheat spoilt by them, so Father decided to grow wheat again on the same field the following year and see if he could not do something to destroy them.

This was just like him, for he would never take the easy way out of trouble but always carried the war into the enemy's camp, even at the risk of further losses.

A farming friend had told him that a green crop of mustard ploughed in was a cure; so as soon as harvest was over he ploughed the wheat stubble and broadcast a good thick sowing of mustard. When this was a good height he ploughed it in, ran the harrow over the field and drilled it at once with winter wheat, following this up with a heavy rolling with a ribbed roller.

How much of the cure was due to the mustard and how much to the roller we never knew, but we had a big crop of wheat the next year. Perhaps the wireworm were so busy with the mustard that they had no time to attack the wheat, or perhaps the mustard killed them as his friend believed.

The oats were sown in the spring on a windy day in March, when the clouds of dust blowing behind the cultivator and harrows showed that there was a good dusty seed bed and it was time to fetch out the drill and get the oats in.

The saying 'a peck of March dust is worth a king's ransom', is a relic of the old days when oats and barley were the common cereal crops and dry ground was essential when they were sown.

Wheat and oats were cut before they were really ripe to prevent losing a lot of grain during the various operations the binder had to do before it dropped the tied sheaf on the ground; they had then to ripen in the shock for some days, in fact many farmers reckoned it took at least ten days before oats were fit to carry.

Barley, on the other hand, was cut dead ripe and carried the same day if possible to ensure getting a good malting sample; but this was generally impossible if a clover crop had been sown with the barley and had left several inches of green clover at the bottom of each sheaf.

Wheat was reckoned fit to cut when no white milk could be squeezed out of a grain; oats when the grain was not hidden inside the leaves of chaff around it; and barley when there were no longer any purple marks on the grain.

The binder used to cut about 10 acres a day if there were no unforeseen hitches; but a binder can only work when the corn is dry, so early morning and late evening work was often impossible owing to the dew, and over-time with the binder was something of a rarity.

Fifty years ago the men always went ahead of the binder and cut a road round the field with the sickle to give a clear run to the binder and its two horses without trampling down any of the wheat. They also cut by hand any badly laid patches of corn and marked off the

strips they had done by tying their own fancy knot in
some straws at each corner; these strips then had to be
measured up with a chain to work out how much each
man got for his piece work.

Setting the sheaves up ten at a time into shocks
looked easy; but there was a knack in it, and badly set
up shocks caused a lot of extra work with unnecessary
reshocking after a high wind. Our men were adept at
it, as befitted workers in a famous arable district, and
nothing but a complete gale would blow down their
shocks.

When carrying began all hands turned to and rushed
the job along as fast as possible during every minute
of the day and often late into the evening.

For many years Father supplied 'harvest beer', and
each man had his own earthenware jar which was filled
from the barrels in the cellar in the early morning and
then taken to the field or the stack and put in the shade.
In later years when all payments in kind were changed
to payment in cash the beer ration disappeared and 'beer
money' took its place.

Father thought this was a step in the wrong direction,
for 'Bavers', or time off for drinking beer, ceased to
have any social significance when it lost the air of family
refreshment that it had if the farmer supplied the beer.
The harvest-home supper and the practice of gleaning
the stubbles dropped out at about the same date; in fact,
it seemed that changing payments in kind for payment
in cash killed that pleasant mutual rejoicing over
harvest which for years had existed between master
and man.

Father was efficient himself at all farming operations
and expected efficiency in his men; he hated slipshod

work, and it was an unwritten law with us that everything on the farm must be done well.

The countryman is supposed to be slow at his work, but this does not take into account the fact that hard work can only be done at a slow rate since the power of a man is limited. He is thought to be stupid by some who have not worked alongside him and tried to match his skill.

But it was corn harvest that always brought out to the full the mastery of the countryman at his work.

Slowly but surely they carried on with the hard work hour after hour of the long harvest day, and the job was done just as well at eight o'clock in the evening as at ten o'clock in the morning. When they knocked off the horses were tired and the men had almost reached the limit of human endurance; yet they finished the last drop of beer and trudged home with a joke over something that had tickled their fancy during the day's work.

One year Father let me work with the men all through hay and corn harvest, and though only a youngster I did spells at almost everything—tedding and horse raking the hay, making silage, pitching corn and loading the wagons and making the stack. The men were always on the look-out to see I did not overstrain myself, and to impress on me that 'slow but sure' was the only way to go about things.

As we all walked up to the rickyard together, following the harvest-home load of wheat, one of them came to me and as spokesman for the rest said they were all proud to have helped the young master—as they called me—to become a good farmer; a very pleasant memory of good country comradeship over forty years ago.

11. The New Arrivals

EVERY September Father went down to Devonshire to see what young Devon bullocks his usual dealer had for him, and to make a selection from these for fattening during the following year. It was a fine combination of business and pleasure for him, and was a good rest after several hard weeks of harvesting.

On his good judgment of the quality of the young beast he bought there depended the success of his next fattening season, the cups and prizes he would win at the shows, the balance in the bank, and the upkeep of his reputation as a producer of fine-quality fat stock.

This reputation was worth a good deal in saving of money and trouble. Our nearest large market town was nine miles away and driving fat beast that distance before sale was a poor business proposition; however slowly and carefully they were driven they were bound to lose condition before they got there.

Owing to his reputation Father seldom had to do this; for the local butchers vied with each other to buy his animals and used to come and walk round the fields with him. They very often bought all the bullocks in a field at so much a stone dead weight, on the understanding that two were delivered to them each week. This was a very pleasant way of doing business on a summer afternoon and was an excellent arrangement; for Father was free to send the ripest bullocks first; and let the others go on fattening until their turn came.

Judging the quality of a finished bullock was fairly easy; most dealers, butchers, auctioneers and farmers could do it at a glance, and the prices fetched in the sale ring by the chosen animals were always the ultimate test of the accuracy of their judgment. But it was not easy to put the make-up of good quality into words.

Perhaps the general bloom on the bullock, suggested by the good condition of his coat, the brightness of his eye and the liveliness of his stance, were the things that made us feel he was of good quality as apart from his weight. This quality had been produced by the excellence of the pasture and the good management of the bullock through a whole year, and these had left an indelible impress on the animal amounting almost to the grazier's trademark.

So definite was this that some butchers used to say that they could pick out Father's entries at the local shows without looking at the catalogue.

But to judge the latent quality in young stock, that so far had seen no good pasture and had probably been living rather rough for some time, was another story. This faculty, given only to few, Father had to the full; but admitted that it was something he could not teach to anyone else.

He said he always pictured in his mind what the young animal would look like after he had fattened it, so no doubt its frame entered into it. He would never take a youngster that was wild in the eye or looked worried, so probably temperament was important. Nor would he ever buy young stock in the market if he could possibly avoid it, as he always like to see them grazing in the open; for a poor grazer never fattened well, and on close inspection generally showed signs of having been underfed at some time in its youth.

Given good breeding, Father always considered malnutrition at some time the main reason why he rejected so many of the youngsters he saw; this poor feeding had set up some chronic disorder, indigestion, lack of appetite, poor bone or bad temper, and no good ever came of trying to make a silk purse out of a sow's ear however tempting the price might be.

When Father got back from Devonshire it was near the end of September and the farming year was nearly dead, but in a week or two Sammy drove up from Roade Station the young Red Devons for wintering and fattening, so the farming year began again.

Sammy was a local celebrity; he was popularly supposed to 'lie rough' always; that was because he did not believe in spending good money on board and lodging, but got his meals of bread and cheese and beer at the various public houses round about and slept in a barn or shed or haystack or even a dry ditch as opportunity offered. In spite of this he was the best cattle drover in the district and never overdrove his charges, lost them, or got them mixed with other people's animals.

At one time Sammy used to have anything that was going for a meal on his arrival. But one day he was given a piece of rabbit pie, and when he brought back the dish Mother asked him what he had done with the bones. 'Bones, ma'am; were there any bones?' he replied. After that Mother decided that bread and cheese was safer.

While Sammy got his five shillings and his glass of beer and bread and cheese the young bullocks were safely housed in a field near the house till Father could look them over, divide them into lots, and decide which fields would suit them best. Meanwhile they were settling down and enjoying a bite of food after many hours on the rail and a three-mile walk from the station.

This dividing into lots was an important factor in the management of bullocks and had to be got as nearly right as possible from the start, so that those which would fatten early and later were kept in separate groups; for once bullocks had got to know each other they did not tolerate changes easily, and a newcomer often got a poor time if it was thrust into a hostile group.

So from the very beginning the groups lived in different fields, and when the time came for them to winter inside they went into different yards.

With us they always started life in the fields on the top of the hill where the ground was drier than elsewhere and the grass not too rich; for young beast must not be started on rich pasture. They were handy here too for the ration of hay, and perhaps swedes and oat chaff mixed, which they were given as the weather got colder. But though they were well fed Father did not believe in coddling young stock, and in mild weather they stayed out until Christmas.

The longer they were outside the quicker they fattened in the yards and when the grass started growing in the spring; but when the weather got so cold that the greater part of the food they got was used up in keeping up their bodily warmth they were brought into the yards, and the real business of fattening them began in earnest.

During the early part of their stay inside they got two good meals a day; but this was increased to three towards spring, for Father found that little and often produced better results as soon as they began to put on weight.

At that time most farmers did not feed cereals to

their bullocks; but Father gave a few pounds of middlings each day to his youngsters in the yards, and so
was able to cut down somewhat their allowance of cake.

12. *Wintering in the Yards*

> Three white frossises allus bring rain;
> If not 'twill be a week I 'low
> Afore 'tis soft again.

WE have had the three white frosts and no rain to
follow, so taking the advice of the local saying manure
carting is in full swing this week in early November; for
several days the whole place will reek with the smell of
manure and the yards and fields will be no place for
anyone who has not a good country nose.

The frost has come very opportunely, for the yards
will be emptied of manure before the young beast want
to come in for the winter; and the arable land will get
its dressing while the ground is hard and the loads
draw easily and make as little mess as possible on the
fields.

As long as the frost holds the manure will lose none
of its goodness and can be spread and ploughed in as
soon as the thaw comes; until then stubbles will be full
of birds, and the blackbirds and thrushes will find plenty
of food in the heaps of manure and will do a bit of
manure spreading on their own.

Father used to run two teams and do two fields at the
same time, using six horses, six men and two boys on
the job, and soon got all the yards cleared at the rate
of about 50 tons of manure a day while the frost lasted.

As soon as the thaw came the men were switched on

to manure spreading and the ploughs turned it in and left the arable land well ridged up. This held the winter rain and let the later frosts crumble down the ridges into a good tilth ready for the spring sowing.

It always surprised Father that some farmers thought they were doing something extraordinary when they put 20 tons of farmyard manure to the acre on a field. The top 9 inches of soil probably weighed over 1000 tons to the acre, and in good fertile loam contained the equivalent of about 100 tons of manure and about 10 tons of basic slag, so that the addition of 20 tons of manure or 10 cwt. of basic slag was nothing to shout about.

Heavy crops of wheat or mangolds, taken completely off a field, removed a good deal of the dry matter of the manure in the soil. As it was important to keep up the store of natural fertility, Father reasoned that the dry matter in the manure supplied should equal that removed by the crop. And he took as his basis of calculation that the dry matter of his farmyard manure amounted to about a quarter of its weight.

As soon as the manure was out of the yards a thorough clean up of them was made with brooms and buckets of water and disinfectant. The brick floors were washed down, the drains opened and flushed, the mangers and racks scrubbed, and the whole place left with a fine healthy smell ready for the newcomers.

Wheat straw for bedding and oat straw for chaff will be wanted in plenty, as soon as the animals come in for the winter, so threshing was always the next item, and the threshing outfit had a standing order to come to us in October.

As this must not be held up by the weather Father had stacked his wheat in one of the bays of the big barn, and the whole outfit except the engine could be run inside under cover. The wheat stack slowly dwindled in one bay and the straw piled up in the opposite bay as the work progressed in an indescribable dust and noise.

The sacks of corn went by the hoist straight up into the granary above one of the bays, and the tailings and bags of chaff into an empty bay so that there was little manhandling to do.

But it took a team of nine; the two men who came with the outfit ran the engine and fed the thresher, two men on the rick threw down the sheaves on to the thresher, one man cut the strings and two made the straw stack, the last two, almost hidden in a cloud of dust, looked after the bags of corn and chaff.

Whenever the hum of the threshing machine was heard it was quite certain that all hands would be at work, and everything else but essential feeding and milking had to be left for the time being.

Father always estimated the quality and wrote down his expected yield of the crop as it stood in the field before cutting. Now as he caught and examined a handful of grain as it poured down the shoot into the sack, and watched the stack diminishing and the sacks mounting up, he knew whether his estimate had been good or only wishful thinking. There was no appeal against the verdict of the threshing machine.

Besides the straw for bedding, chaff and other feeding stuffs such as hay, mangolds and cake were now collected into the barn; for it was connected by covered ways to the milking sheds, the stables, and the fattening yards—in fact to every place where food had to be taken daily.

It was a great advantage to the cowman, wagoner, and cattleman in bad weather in winter to have all their

feeding stuffs handy and water laid on; they had not to
go outside for anything. The covered ways were lit
by lanterns hung in sheltered places so that the work
could go on in the early mornings and evenings without
one hand being used to carry a lantern.

This was the season for our annual rat hunt so that
we should not lose large quantities of the feeding stuffs
collected in the barn. During the day the rats used the
rickyard, but every evening, when all was quiet, they
went into the barn for warmth and food and shelter for
the night. So late in the evening before the rat hunt
we went round the outside of the barn and plugged all
the holes that the rats used with sacks or pieces of wood.

Next morning with trousers tucked into our socks
we went in with Spot, our white wire-haired terrier,
and when we had opened up the top half only of the big
doors for some light the hunt began.

Spot was a born ratter and an expert at his job; he
seized the rat by the back of the neck, gave it one bite
and a shake and then flung it away ready for the next
victim.

A few unwary rats were soon disposed of, and then
began the real hunt amongst the bags and implements;
as we moved each bag Spot stood by tense and still until
a rat moved when in a flash he had it; we turned the
chaff cutter and one dropped out into his mouth.

Towards the end the few survivors tried to climb the
walls and managed to get a few feet up in a corner; but
it was a vain effort, for Spot stood below and waited
for the inevitable moment when the rat would lose its
hold and drop down to its death.

A bag of about fifty was considered good on the first
day; then the holes were opened up again and in a few

days the hunt was repeated, but this time Spot was generally disappointed, for we seldom accounted for more than a dozen.

To carry on the good work all through the winter we had an excellent trap made out of a big galvanised iron bin half full of water. A spindle carrying a tilting lid was put in the top and a red herring hung on the far side; a piece of wood acted as ladder to let the rat climb on to the lid, then as he crossed over to reach the red herring the lid tilted and he went into the water below. The lid turned slowly back by means of a small counterweight and the trap was set again ready for the next victim.

One bait lasted for days, and the older it got the more attractive it became to the rats; if the balance of the lid on the spindle was good, and the spindle was kept well oiled, a very small counterweight was necessary, and a night seldom went by without a bag of some sort in the trap.

When all this had been done and the yards were clean, cleared of pests and vermin, and provisoned with everything that would be wanted, then the young bullocks could come in at any time as soon as the grass got short or the weather too cold.

Long after the ordinary grasses and clovers had ceased growing the rye grass kept on, with the result that it got overgrazed and weakened for the following year if the bullocks stayed on it too long; also in cold weather most of the food of the animal kept in the open went towards keeping up its bodily heat and left none over to fatten it. So for the sake of the animals and the pastures it was best to take the young stock in about Christmas time.

In the covered yards, bedded down each day with fresh straw and kept dry and warm and out of the wind, a good proportion of their food went to fatten them; and the food they got might be thought of as a maintenance ration to keep them fit and growing, and a production ration to put on flesh and fat.

The change from an active life hunting their food to a lazy one, with meals brought to them twice a day, often had the effect of making their appetites not so hearty as when they were out of doors. Until they had got accustomed to their new conditions the cattleman had to be on the watch to see that they did not go off their food; this he did by varying their diet and by making them eat chaffed oat straw by mixing it well with sliced mangolds which they liked.

A typical daily allowance for each two-year old bullock was:

> 40 lb. sliced mangolds
> 10 lb. best hay or silage
> 10 lb. chaffed oat straw
> 4 lb. middlings
> 2 lb. cotton cake.

Of this we might consider the cotton cake and middlings as the fattening ration, all the rest being the maintenance ration.

If the mangolds, hay, middlings and oat straw had to be bought this would work out at about 1s. 6d. a day for a winter of about 100 days, or £7. 10s. per head. Obviously a bullock would not appreciate in value by this amount in that time, so that it was essential that most of the food was home-grown and not bought foodstuffs.

13. Horses

FIFTY years ago the standing of a farm could be judged by its horses and wagons, and we had eight shire horses, which were lovely animals and took many prizes.

When they were ready to start for a show they always paraded in the courtyard for Mother and all of us to see. They were a magnificent sight with their coats shining with health and much grooming, their manes and tails plaited with gaily coloured ribbons, and their polished brasses flashing in the sunlight; in the evening on their return we all went out again to find out which carried their prize rosettes.

Father enjoyed everything to do with horses and bullocks, but he hated cows because they were such hard taskmasters and made a terrible mess of the place. When the opportunity came to rent a dairy farm on the other side of Stoke Bruerne, and much nearer the station, he transferred all the cows there with great satisfaction and so made much more room in the yards at home for his bullocks.

Father was a good judge of a horse and a fine horseman, and in his younger days hunted twice a week with the Grafton; he taught us all to ride almost as soon as we could walk, and schooled us over jumps to such good purpose that in later days my brother and I used to break in our own young horses.

For farm work he had his favourite mare Kitty; she was almost as good as a dog for handling sheep, knew how to come up to every gate on the farm for easy opening, and had been trained to stay still as soon as the reins were dropped on her neck even when her rider dismounted and left her.

One day when I was riding Kitty a bad thunderstorm came on and we sheltered during the worst of it under a hovel. A very loud peal of thunder shook the hovel and so frightened Kitty that she bolted for the open hitting me against one of the pillars of the hovel and brushing me out of the saddle. I landed heavily on the ground and for a moment or two was stunned; but when I recovered I found that Kitty's training had held in spite of her terror, and that she was standing a few yards off me with the reins on her neck and looking round at me to see what was the matter.

She was just as handy when driven in the dogcart, and never put a foot wrong on ice-covered roads in winter, for she would put her feet together and slide until she could get a fresh grip.

Two chestnut brown shire horses made a splendid plough team, and they and the wagoner did some lovely work at ploughing matches.

How simple yet how efficient is the plough; the coulter cuts the vertical slice, the share set at anything from 4 to 9 inches deep cuts the horizontal slice, and the mould board gradually inverts the solid rectangle of earth cut by the coulter and the share as we move along.

And the ploughman has the company of his horses, who know their job just as well as he does, and the birds which continually wheel overhead and settle behind him as he goes.

To see a team of three horses and a plough turn at the end of a furrow leaving only a short unploughed headland of about 15 feet is to realise the perfect co-operation of man and horses in the doing of a perfect job.

No wonder ploughing matches used to be a regular feature of the arable lands; for in these contests not only had the ploughman to be first class, but he had to have this perfect co-operation between himself and his team to have any chance of winning.

It will be a sad day when competitions like ploughing matches and sheep dog trials die out.

Every year after the threshing the wagoner and his boy loaded up the sacks of corn for sale on a wagon and set off with them to the mill.

This by country custom was the wagoner's holiday; he set off in his best clothes, with his best wagon smartened up for the occasion, and his best team of horses. Very often a new coat of paint had been given to the wagon, new brasses bought for the horses and a new whip for the boy, all out of the wagoner's own pocket.

For there was a proper pride to be shown on the road, pride in the wagon to show it came from a good farm, a personal pride in the team and their accoutrements, as well as pride in the sacks of good wheat from the fields where he worked.

He would get back late in the afternoon after a happy day with his horses away from the humdrum work of the farm, and had much to tell of those he had met on the road and the interest taken in his own team at the public house where he and his horses had their midday meal.

Heavy loads of hay or corn on big wagons took a bit of manœuvring, and gate posts were liable to be carried away unless the boy leading the load had some common sense. The village boys soon learnt under Father's eagle eye to take a gateway properly; one in particular was given an object lesson that he never forgot.

Father found him bringing a load drawn by Bonny, one of the cleverest of our shire mares, up to a gateway at much too fine an angle to get through safely. So he stopped him and turned the wagon back into the field and brought Bonny up towards the gate again, leaving her some distance from it. Then Father joined the boy at the gate and told him to watch how Bonny knew better than he did how to negotiate a gateway.

With a 'gee up Bonny' the demonstration began. The mare made no mistake, for she brought the load well across the field until level with the gate, then turned squarely on to it, and came through triumphantly without any leading and turned up the road to the rick-yard. As the road was uphill she stopped in a few yards to wait for the boy to hook on the trace horse to help pull up the hill. There was no doubt that Bonny knew better than the boy how to handle a heavy wagon.

Horses are very destructive to a good pasture; they cut it up badly and their habit of grazing makes the management of the grass well-nigh impossible, so that they are never allowed on the best fields.

Every animal has a different method of grazing according to the formation of its mouth; cattle and sheep have no biting teeth in the top jaw but a hard pad takes their place, so they more or less pull off the grass and find really close grazing uncomfortable; but a horse has biting teeth in both jaws and can nip off the grass quite close to the ground.

Horses are fastidious eaters and gnaw down the grass on the part of the field they fancy almost to the ground, biting off all the runners of the clovers and grasses so that the bare earth shows and the sward is weakened in hot weather; to make matters worse they always leave their droppings on the part of the field they do not like.

Thus the good parts of the pasture are overgrazed and the poorer parts left rank, which is the worst possible treatment for grass. The only remedy is to fence off the overgrazed portion and so force the horses to clear up the parts they do not like and which now have grown rough. A little salt thrown over the rough stuff encourages them to eat it off, but even then they will only make a poor job of it. After they have done what they will it is best to give them a change of quarters, spread their droppings, and then put sheep and beast in to tidy up the pasture.

All our horses spent the summer out at grass when not at work. In the winter they were watered, fed and groomed in the stables, and then put in a covered yard for the night or for week-ends so that they could move about and get some exercise.

14. Sheep

MANY a time Father expressed a wish that nature had made an animal as useful as a sheep without all its disabilities.

An animal in fact that did not have its young born about midnight, that did not get cast and die on its back unless found, did not get foot-rot or need washing, dipping, shearing and hunting all over for maggots; and as a final afterthought he would add, 'and not so darned difficult to count'.

But in spite of all their bad points every grazier had to have sheep. No other animal could take their place for getting a new pasture into good order; for treading down and consolidating and manuring a light soil when folded upon it and so getting it ready for corn in the following year; for scavenging the tops of the mangolds and doing well on the scavengings; and perhaps best of all for putting in amongst bullocks to help eat off the grass in good grass years and so keeping the pasture in proper order.

For when the grass outgrew the rate of grazing of the bullocks in a field it was impossible to add more bullocks to help graze it down, for the old hands would fight with the newcomers. But bullocks never worried about sheep and tolerated their presence with an easy contempt that was amusing to see.

Yet in spite of their many failings our old shepherd thought the world of his sheep; he was a grand old man, with a face like a Bible picture, and he never grudged the weeks he had to stay away from home at night or the many midnight hours in the lambing pen when the lambing season was on.

'They ain't made no machine to run lambing time, mister, nor never wull, praise be', he used to say with obvious pride when we went round to his hut, close to the lambing pen, to have a chat with him in the evening and to hear how things were going. 'And they ain't made no mechanical dog yet neither, have they Bob', he would add, as Bob came from his corner and laid his head across his master's knees.

He still used to lead his sheep, with Bob bringing up the rear behind the flock, when they moved from one field to another; and this was a sight that always astonished any townsman.

Horses and rabbits are biters and gnawers and can do a lot of damage to pasture and trees, but cattle and sheep are true grazers and ideal travelling manure carts, as they leave their droppings evenly over the field. Their cloven hooves act as cultivators of the soil; and sheep, owing to their lighter weight, can go on grazing a field in wet weather after the cattle have had to be removed for fear of damage by treading.

There is nothing like a herd of bullocks to take off the rough stuff, and a flock of sheep to eat the grass down after the bullocks have been over it, to give a pasture that complete grazing which is so good for it once every year.

No farmer can keep his pastures in good condition by specialized grazing by one kind of animal only, and Father believed that more grass was ruined by bad management of the grazing than by any other cause.

In his opinion a judicious combination of grazing by bullocks and sheep made it possible to improve an old pasture; while no new pasture would ever become worthy of the name unless in its early days it benefited from the treading and on and off grazing of a large number of sheep before any bullocks were put on it at all.

Sheep have the additional advantage that they can clean up a pasture after the grass begins to fail in the autumn and the show bullocks have been taken into the yards. When the pasture needs its winter rest they get their living on the arable until lambing time in the next spring.

Only at lambing time do they cease to be self-supporting, and have to draw on the supplies of food in the larder for their hay and cake that they get in the lambing pen.

Although a flock of sheep spells continuous hard work for the shepherd it also brings good heart to the pastures of every grazier who aims at producing good-quality fat cattle. The golden hoof not only brings corn but good grass in its wake.

15. Repairs

'SHOW me a tumbledown farm and I will show you a poor farmer', was Father's way of saying that good farming and good repair work went together.

As he was always determined to keep everything in good order he was never ashamed to show any visitor round the farm; for the whole farm had an air of prosperity, and that look which a place has when a proper pride is taken in it.

The big jobs of hedging, ditching, draining and cleaning out the drinking places at the brooks, as well as the smaller ones of mending gaps, renewing gates and gateposts, and repairing roads and buildings were done each year.

Ours was a hunting country and we were only a couple of miles from the Grafton Kennels; so that gaps in fences could not be mended here and there with odd bits of wire but must be stopped properly with posts and rails. Gates must be made to open and shut easily or they would always be left open, fences that must be protected by barbed wire had to be marked with the proper red warning sign, for Father hunted as well as farmed and had first-hand experience of both the farmer's and the hunter's point of view.

Gravel had to be drawn from the local gravel pits to mend the roads, which showed signs of the hard wear they got in hay and corn harvest; the edges were trimmed up so that they looked like roads and not grass rides.

But these were minor repairs, and the jobs that took the most time were hedging, ditching and cleaning out the drinking places.

There were two kinds of hedges on the farm; the low hedges that were trimmed quite frequently and the tall hedges that were tackled only about every ten years. But some of these tall hedges had to be done every winter; if they were left too long they got very thin at the bottom and the animals crept through.

Attacking a hedge of this sort was a formidable business and an art in itself, and the hedger fetched out his heavy gloves and his slasher, billhook, and baghook before he set to work to 'lay' such a hedge.

First of all he slashed off a lot of the straggling side growth and brambles to make room for him to work at the bottom of the fence. Then with his billhook he cut out all the dead wood and as much of the live wood as he did not need; by this time the hedge looked as if nothing could be done to make it good again.

But some of the live wood was now cut into posts of the right height for the new layered hedge, and when these were driven in and the stumps of others cut to the same height he had a framework on which he wove the long straggling branches. This he did by nicking them halfway through close to the ground, bending them over and twining them in and out between the posts and stumps. Then with his baghook he pruned off all the odd bits and shoots so as to leave a perfectly interlaced fence strong enough to keep in any cattle and good enough to be a match for any hedge in Northampton-shire, which was saying a good deal.

For a month or two it looked rather bare, but when spring came it was soon a mass of new shoots, and in a short time was hardly distinguishable from the other low hedges on the farm.

Drainage was expensive and ditching comparatively cheap; so it was the worst folly to do away with the advantage of good drains by having blocked-up ditches, with the result that water stood in the drains some way up the field, the manure put on was largely wasted, and the land began to turn sour.

As the earth taken from the ditch was an excellent top dressing it was always spread on the grass fields, and gave some little return to put against the expense of ditching.

Many farmers neglected draining and ditching because they did not understand their importance, or did not feel competent to deal with defective drains.

Father always considered they were the keystone of good farming and would always examine them first on any farm he went to inspect. So long as they were not in working order it was just throwing money away to put good manure on the arable land or to attempt to fatten good stock on the pastures.

For a gradual and increasing deterioration set in on both which could not be checked. Eventually when the land had deteriorated so badly that the farmer saw ruin staring him in the face, the remedy cost much more than he could afford; whereas if the rot had been tackled at the outset, everything could have been put right for a few pounds.

The provision of drinking water for cattle on a grazing farm is often a serious difficulty, for a cow may drink as much as 20 gallons and a bullock 10 gallons a day. Those farms like ours which had good meadows, with plenty of drinking places, tended to keep them as permanent pastures for this reason; for the water in the other grass fields had to be laid on by pipe or even carted to the drinking troughs.

Drinking places in the brooks must of necessity be rather elaborate and needed a lot of attention; posts and rails round them must be kept in order or the bullocks would break through and after walking up or down the stream get out into another field.

The approaches must not be steep and must be made of hard stone and not just mud, or the drinking place would soon be blocked up; and allowance must be made for the different level of the water in winter and summer so that the drinking place was still usable in a drought.

A farmer might lose a good bullock through careless-ness over these points. When the proper drinking place was out of action for any reason the bullocks would certainly get into the brook somewhere else. If one got bogged, and could not get out, it might die before it was found.

16. Spring Cleaning the Pastures

SOME farmers' only idea of spring cleaning a pasture is to harrow it; this looks nice but should be only the last of a series of operations all intended to put the grass in good trim after its winter rest.

Father's spring cleaning did not all take place in the spring, and the chain harrowing was but the final bit of spit and polish to round off what was often a lengthy process of rejuvenation. For all plants must have air, water, food, warmth, root room and a soil that is not sour, and grass needs all these just as much as the arable crops.

The ordinary methods of cultivation of the arable land provide all these without much difficulty, but it is much more of a problem to provide them all for a grass field that we cannot plough up.

Except for looking after the drainage, our role must be that of encouraging the minor ploughmen and tunnellers that live in the soil to do the work for us.

These humble engineers—wireworm, leatherjackets, worms and moles—prune the grass roots, leave air pipes large and small, throw up earth to the surface which when spread makes an excellent top dressing; in fact, they spend their time in providing raw material for the humus, making conduits for the rain to follow, and air conditioning the soil.

Above ground rooks and starlings hunt for these same humble engineers; they probe and loosen the surface and are an indication that all is well below ground.

Father knew it was wise to investigate what was happening when worm casts and mole heaps were not seen in any field and the rooks and starlings shunned it. He generally found that such consolidation of the surface soil had taken place that it no longer breathed properly nor could the rain penetrate easily; and that in consequence the underground miners had gone on strike.

Then the only remedy was a good dressing of lime to loosen the soil and do some of the work that the absent miners had done before they went on strike; after that they would probably return to work in a month or two, and all would be well again.

An open soil such as encouraged wireworm, leatherjackets, worms and moles was more airy, had more available moisture, more root room, and was warmer and richer in plant food than a consolidated soil.

But some of the mineral constituents might still be missing, and other indicators must be consulted to test this.

Father kept a lump of rock salt as a salt lick in every pasture, and if the bullocks made a lot of use of this he took it as an indication of lack of potash, and a dressing of kainit was given; similarly, a lack of white clover showed a need of phosphates, to be remedied by a dressing of basic slag; while if sorrel grew too freely the pasture was getting sour, and a dressing of lime was indicated.

Thus every pasture had its careful inspection and scheduled treatment for its spring cleaning; and all this was done before the chain harrow scattered the animals' droppings and the mole heaps and gave it that lovely brushed coat of light and dark green which was always an infallible sign of spring.

Nor did it end there; for in May the buttercup heads were mown down, and after haymaking the rough un-grazed patches in the pastures were cut over with the scythe on a sunny day, and the bit of hay so made was soon eaten up by the bullocks.

In a few days they would return and graze the after-math closely, so that the field was grazed all over by the end of the summer; later on the thistles and docks were cut down before they could seed and the animals' droppings spread at the same time.

Obviously all this cost money and too many farmers grudged this and so never arrived at having pastures worthy of the name.

Father knew that a pasture in our part of the country was only doing its work if it would fatten at least a bullock to the acre without cake and would put on about

200 lb. of dead weight to each bullock in the fattening period. It was only the constant inspection of the sward coupled with the progress of the animals grazing it that enabled him to keep up his high standard of fat beast.

Often enough those who came to look round the farm would say, 'I wish I had grass like this'; to which Father would reply, 'And so you could if you took as much trouble over it as I do'; and though he said 'trouble' he did not really mean it, for to him it was not trouble but a real labour of love.

The money he spent on cake for his bullocks and on the general upkeep of the pastures was a good investment and repaid itself. But I believe he would have spent it anyway even if he saw little or no return for it, for keeping his grassland in good heart was the first article of his farming creed.

Of course the whole of the grassland was not of this high quality, and on the lighter soil on the high ground near the farm buildings the difference was quite noticeable. This poorer quality grass was used for young stock that did not need first-class grass, for the sheep and lambs when not folded on the arable land, and for the horses which got extra rations in the stable when at work.

But this poorer grass was treated well too, for young stock and lambs will assuredly make the poorer pastures poorer still if a dressing of basic slag is not given to replace the lime and phosphate they remove from the soil for their growing bones.

To get an 'early bite' in any field it was essential that the really early grasses should be present in the sward, and that they should be encouraged to grow in what was not really proper growing weather. Father found that the easiest way was to flood one of three meadows each year, so that the early bite was only taken off any one meadow once in three years.

The River Tove split into two branches and watered three of our meadows, and by dropping railway sleepers into fixed grooves in front of the arch of a brick bridge over the stream any one of these meadows could be flooded at will by means of the various trenches that had been dug for that purpose.

The water was put on in February for about a fortnight and timed to come off again in that short spell of warm weather which was so usual with us about the end of February or the beginning of March. If everything went well the grass sprang up as if by magic and a splendid early bite was secured.

Normally we had no gulls following the plough as we were too far from the sea; but the water was never out more than a day or two before they appeared, and from then on the meadow became a feeding ground for gulls, herons, duck, geese and terns on occasion.

Father always thought that the wonderful way the grass grew after flooding was not due merely to the water, as floods earlier in the winter never produced the same result, but to the goodness of the arable land that ran away in the drains and so to the brook; and that he tapped in the brook not only some of his own wasting goodness but also that of all his neighbours upstream.

17. Summer in the Pastures

THE different animals reacted very differently to their first sight of grass after their long winter indoors. Horses after kicking up their heels for a few moments found a place to have the luxury of a good roll; the

more stolid cows could hardly wait to get through the
gate before they began to eat the fresh grass; while
bullocks wandered all round the new pasture and
examined everything before they settled down to graze.

The choice of field for any group of bullocks that had
been kept together all winter in a yard was not a mere
matter of chance. Father worked out a complete grassland rotation and had prepared three fields for this
group, so that grazing should be continuous and no
unnecessary setbacks should spoil the fattening process.
This is best shown by the following table:

	FIELD A	FIELD B	FIELD C
1st year	Farmyard manure in autumn Hay Aftermath for grazing	Basic slag in autumn — Summer grazing	Early bite Rest Late bite
2nd year	Early bite Rest Late bite	Farmyard manure in autumn Hay Aftermath for grazing	Basic slag in autumn — Summer grazing
3rd year	Basic slag in autumn — Summer grazing	Early bite Rest Late bite	Farmyard manure in autumn Hay Aftermath for grazing

Field A was dressed with farmyard manure in the
previous autumn, gave an early crop of hay in the 1st
year and so had a good aftermath ready for the bullocks.
Some of the manurial value of this farmyard manure
persisted into the 2nd year and provided the early bite

and after a rest the late bite. A dressing of basic slag in the autumn produced a good increase of white clover for summer grazing in the 3rd year.

Similar treatment was given to fields B and C. The order in which the fields were used by the bullocks in the 1st year was C, B, A, C.

In this way each field had its appropriate manuring and a rest from grazing during the season, and the bullocks enjoyed fresh grass at each move so that there was no setback in the fattening process.

Our bullocks got cake every day and it was essential that each got his proper share; this was not so easy as it sounds for they loved cake, and rather like children were not above grabbing a double ration if they could get it.

We found that the best way was to put the cake into iron cake pans—one for each bullock—which were well separated in a fairly big circle. When the cattleman appeared with the bag of cake the bullocks soon learnt to come to the pans, as he kicked the first pan and so rang the dinner gong which sounded all over the field.

The cattleman went round the circle putting a half ration in each pan and arrived back at the first pan he had filled just about the time the bullock had finished his first helping. Then he gave out the second helpings and in this way encouraged the bullocks to stay at their pans.

If a greedy bullock ate his own share quickly, and then wandered round the circle trying to steal another ration, the cattleman prevented this; and it generally only took a few days to show any greedy ones that nothing was to be gained by running round.

After that every bullock stood at his pan and waited for his rations to arrive.

Father loved the summer days when he could ride round the farm on Kitty and inspect all his animals; the sheep, horses and cows were seen daily by the shepherd, wagoner and cowman, but Father always made a daily inspection of his bullocks himself. By constantly being amongst them he got to know all about them and they became so accustomed to him that most of them allowed him to handle them whenever he liked.

He soon spotted any nervous beast that grazed by itself and was slow to come up to the cake pans; if this nervousness did not wear off he put all the nervous ones from each group together in one field, and it was wonderful what a difference this made to them. In a few weeks time bullocks that had not got on in their own groups settled down into happy contented animals with well-licked coats and filled-out bodies.

He was always on the look-out for any bullock that was out of sorts, and by catching it early and drenching it saved more lasting trouble. A well-stocked medicine chest was kept at home for these minor ailments, and we seldom had to call in a veterinary surgeon.

As the summer went on Father began to choose out those bullocks that he would keep for the Christmas shows at the end of the fattening season on the grass. Here both quality and weight counted, but as he was an excellent judge of quality and had the most uncanny knowledge of the weight of an animal he seldom made a mistake.

The chosen bullocks were left in their group, but when the time came to sell they were not included in the lot for sale. It was always a recognised joke when the butcher came round with Father to buy the bullocks that he asked which he could not have now, and would have to pay more money for if he wanted it after the show when it wore its prize rosette.

By the end of the season Father knew his bullocks so well that it was like parting with old friends when the time came for the inevitable journey to the butcher; but he still had his show entries to take an interest in and soon there would be a fresh lot of youngsters to take the place of those that had gone.

As soon as the grass showed signs of failing in the autumn those chosen for the shows were brought into the buildings, but this time each had a loose box to itself and became a pampered animal.

For a few weeks they were the pride of the cattleman, who spent a lot of time on grooming them and feeding them four times a day. To make them put on fat and get a good gloss on their coats they had extra linseed cake or any food they fancied.

These were the days when big joints of fat beef were in fashion and prize bullocks were really weighty.

By the time the day of the show came they were much too heavy to travel on foot and they were taken to the show in a large cattle float; food was taken with them and the cattleman went in charge of them to get them put into their right classes. As soon as they were settled down they were fed and given a last grooming, so that they should look their best and show themselves off well before the judges.

It was a great day for the cattleman, and if all went well, and his charges got the prizes he hoped for, he felt well rewarded for all his work with them during the last twelve months.

18. Fat Stock

FATHER had many good years and won numerous cups at Towcester and Northampton fat-stock shows, but of all these one memorable year sticks in my memory.

It had been a lucky year all through, for he had picked up the best batch of youngsters we ever had, and the weather had been mild for them right up to the New Year.

When they had come into the yards they had started fattening at once. An early spring and good grass all the summer had helped them along, and by the time the rest were sold we had four outstanding bullocks ready to train on for show work.

When, just before Christmas, they were loaded up to go to Northampton show, they looked as likely winners as any we had ever sent.

When we arrived at the show we had a look round to see what we had to compete against, and Father's verdict was that there was nothing to touch our four amongst the rest of the exhibits.

So we went to have some lunch while the judging took place for Father hated being about when this was going on.

On our return to the showyard we met our cattleman who was very excited and blurted out, 'They've all done it, master.' It seemed too good to be true, but we found that he was right, and Father had won three cups and an extra prize of £30 for the four best beast in the show.

As we drove home late in the afternoon Father was very silent as he often was after an exciting day; but as we came up the hill to home he said, 'It is nice to have done it once, for that sort of thing only happens once in a lifetime.'

He was right; for although he won many more cups, that was his peak year and he never repeated this wonderful performance.

19. Keeping the Land in Good Heart

WHAT exactly do farmers mean when they say they believe in keeping their land in good heart?

Father considered land was in good heart if it would produce, without undue expenditure, crops and meat suited to it, which were of good quality, free from disease, and as abundant as nature allowed.

Thus producing good heart in his land was within the power of every farmer according to his circumstances. There might well be more virtue in improving a poor hill pasture, so that it would feed one more sheep to the acre, than in keeping up the fertility of a naturally fertile farm.

Farmers cannot change the geological structure of their land or the weather, and without undue expense they can only change their soil within narrow limits by draining, liming, and manuring. On the whole they have to work with nature and grow the crops that suit the soil and the weather. Striving after abundance by forcing an unwilling crop to grow in an antagonistic soil or an unsuitable climate is almost certain to be a failure.

It is easier to improve the good heart of a poor farm than a good one. The nearer a farm approaches to ideal conditions the more thought and ingenuity it requires to improve its good heart. It is proverbially difficult to gild the lily.

But whether nature be kind or unkind every farmer can produce a certain number of crops of good quality and free from disease. Good-quality disease-free crops mean healthy livestock, a matter of no small concern to the farmer.

It is easier and cheaper to keep the land in good heart by efficient grazing, cultivation, and manuring, than to combat the weeds, pests, and diseases that thrive when these are neglected. The battle against pests may be won and the crops saved, but quality is apt to disappear in the process, and battles of all kinds are notoriously expensive.

Fifty years ago farmers would have denied that they were geologists, physicists, or meteorologists. But the basic principles of all these sciences were in them and were embodied in their farming tradition. Handed down from generation to generation and continually improved by each farmer's successes and failures, this tradition was the working law by which they kept their farms in good heart.

In this scientific age it is the fashion to debunk everything not scientific and to deny the value of farming tradition. Yet it is well to remember that soil physics is a very young science, but farming tradition is very old. What may be possible at an experimental station, where money is no object, is manifestly impossible on a farm where every experiment must be paid for out of the farmer's own pocket.

Experiments made under controlled conditions are likely to produce quite different results when repeated on a particular field on a farm where the conditions are different.

Good farmers do not decry science or the help it can give, but they realise that its results must be applied within the limits of their pockets, with commonsense, and in the light of their own intimate knowledge of the peculiarities of every one of their fields.

In fact, science and farming tradition must work hand in hand.

Yet present-day conditions demand a radical change in this farming tradition owing to the shortage of farmyard manure. An age-old tradition, based on long experience, cannot be changed in a night. Farmers know that the well-trodden roads are the safest and that pioneering on a large scale is a gamble; and gamblers must be prepared for losses. They know too that no single solution for the upkeep of the good heart of the land without plenty of farmyard manure can be applied on all soils and in all climates; adjustments will have to be made by trial and error to suit the different conditions in different districts.

Whether the solution lies in ley farming to produce the goodness for the following arable crops, or some other system, it seems certain that the bullock has got to come back into the picture. If the plough has to go round the farm, the bullock as travelling manure cart must follow it up, and good-quality beef produced on good-quality grass must be made a paying proposition once more.

If those farmers who can afford to make plenty of trials—and errors no doubt—will volunteer to try out any suggested schemes in their own districts and invite their fellow-farmers to see the results, they will be doing a magnificent public service, and will be forming that new farming tradition which must come to keep the land of Britain in good heart.

Had Father been alive he would have embraced this opportunity with both hands, for he had found out many years ago at Shutlanger what system of ley farming was best suited to his own needs. The names of the fields involved—the Leys, the Middle Leys, the Far Leys, etc. —still bear record to his experiments, and the relative merits of ley farming and permanent pasture were given a fair trial.

In our hot and dry summers a nurse crop for the seeds was essential and the progress of the ley was slow. This made it inevitable that the ley was unremunerative for the first year or two; after that it gave good early and late grazing and would fatten a greater number of bullocks per acre than old grass. But Father always doubted if it gave as good a return as a properly looked-after permanent pasture, unless it was left for at least seven years before it was ploughed up again.

So he concentrated on short leys of clover and lucerne for hay or silage and a luscious aftermath for grazing, and then put the plough in again after one or two years. Both clover and lucerne did very well with us and he got large crops of winter feed for his Devons, and as the ploughed-up ley needed no extra manure he was able to give extra farmyard manure to his root crops.

20. 'Tout Casse, Tout Lasse, Tout Passe'

OUR neighbour at Stoke Park had this motto in letters of stone as a fence to his rose garden. How true of farming as well as rose growing; if the good heart of the farm is not kept up by continual hard work and good cultivation it is not long before change and decay and ruin set in.

Father left Shutlanger Grove in 1904, and after that it fell on evil days. The breaking up of the estate of which the farm was part, the general agricultural depression after the war of 1914–18, and some bad farming combined to leave it in such a poor state that Father was saddened to see a good farm and an old friend so ill-used as he found when he revisited it after over 20 years.

The roads were grass rides, the gates had fallen to pieces, the hedges were uncut and the drains and ditches were blocked. The arable land was sour and foul with couch and weeds, the pastures were full of ragwort and thistles and the clover had almost disappeared.

All the good heart had gone out of the land, and the men had mostly drifted away to other work on the roads and railway or in the ironstone quarries.

Father never went back, as he said he could not bear to see it again so fallen from its high estate.

But, if he could know, it would gladden him to hear that since 1930 the present tenant and his sons, by continuous hard work, have restored the farm to its old fertility and vigour.

Once again the arable land is clean and well manured and producing good crops, and good-quality cattle are fattening on the old pastures.

And so, 40 years after Father left it, Shutlanger Grove has returned to that state of good heart which it should never have been allowed to lose.

ited in the United States
Bookmasters